CLIMATE CHANGE: CHALLENGES AND OPPORTUNITIES FOR UNITED STATES PACIFIC COMMAND

> Of the issues where long-term thinking might drive the need to act differently now, climate change and energy is the plainest, for absent a vision of the future, current policy might do very little. Dramatic reductions in green house gas emissions will not stop global warming; too much already has accumulated. This means that the United States will need to focus as much on adapting as on reducing emissions—a significant departure from current policy.
>
> —Gregory F. Treverton,
> RAND Corporation

The Strategic Context

Any discussion of the impacts of climate change must begin with a discussion of the science behind it. Disagreements in the scientific community concerning the causes of climate change, potential impacts, and possible remedies to mitigate its effects drive diverse public opinion in the United States. Many people have fatalistically concluded that climate change will result in irreversible environmental disaster; however, a large population segment also discounts most or all of the science behind climate change as either flawed or tainted by political motivations. The great majority of Americans fall somewhere in between, convinced that some aspects of climate change theory must have merit, but skeptical of the dire warnings, feeling they are buoyed by political motivations and agendas. Even the terms involved in the discussion cause confusion: are we talking about climate change or global warming? Are we talking about manmade effects on the environment caused by carbon emissions, or natural shifts in the earth's climate that have occurred throughout history?

This paper purposefully avoids a "deep dive" into climate change science, instead starting with the assertion that climate change poses a serious threat to

America's national security. A general consensus exists within the scientific community on climate change and the trends that can be projected into the future.[1] The non-profit Center for Naval Analysis summarized this consensus in 2007 by drawing information from the Intergovernmental Panel on Climate Change, peer reviewed scientific literature, and other government sources.[2] They derived four foundational conclusions. First, while natural forces have always influenced the earth's climate, human-induced changes in levels of greenhouse gases play an increasingly dominant role. Second, the increase in the average global temperatures over the last half century can be attributed to human activities with a certainty of more that ninety percent. Third, those temperature increases have already affected various natural systems in many global regions. Finally, future changes to the climate are inevitable.[3]

Thus the United States faces a strategic challenge from climate change necessitating both strategies to reduce the human activities driving climate change and adaptive strategies to deal with inevitable environmental impacts that will occur. According to RAND's Gregory Treverton, this means the United States will need to focus as much on adapting as on reducing emissions—a significant departure from current policy.[4] Unfortunately, debate on how to deal with climate change remains largely mired in the preventative realm, dominated by extreme positions on both ends of the debate. Partisan divides over the feasibility of actions to reduce greenhouse gas emissions ignore adaptive strategies needing immediate development. Current worldwide economic recessionary fears, high unemployment in the United States, and governmental debt-reduction activities make any actions or regulations that could negatively impact American businesses and risk further economic downturn politically untenable. Economic conditions make it unlikely the United States will sign-on any time soon to international strategies to prevent climate change, and the distant, ill-defined benefits inherent in any preventative strategy unfortunately cause collateral damage on adaptive strategies. The Obama Administration's early efforts towards making climate change a top national strategy priority have stalled significantly.

In her in-depth analysis of the national security implications of climate change, Carolyn Pumphrey asserts that adaptation involves finding ways to accommodate ourselves to what is going to happen.[5] With climate change, there are no certainties. However, vulnerabilities based upon the most likely impacts of climate change are clear. The predicted effects of climate change over the coming decades include extreme weather events, drought, flooding, sea level rise, retreating glaciers, habitat shifts, and the increased spread of life-threatening diseases.[6] Adaptive strategies to mitigate these effects would reduce political instability and sustain global markets. Stable governments and markets represent key strategic interests of the United States.

A key distinction exists between adaptive and reactive strategies. In finding ways to accommodate future events, adaptive strategies prescribe prioritized investment of resources synergized with careful strategic analysis of the future to deal with potential challenges and seize opportunities. Conversely, a reactive posture risks delays in addressing the impacts of climate change until they reach a threshold at which options to mitigate environmental phenomenon result in limited and extremely costly choices. Creating feasible, acceptable, and suitable strategies to face climate change involves finding opportunities for the United States and its partners to appropriately mitigate risk without ignoring other national security concerns. In addition, humanitarian assistance capabilities serve American interests whether needed as a result of the worst fears of climate change being realized or as a result of events not related to climate change.

Climate change threatens to act as an instability accelerant or "straw that breaks the camel's back" in many unstable regions of the world, likely providing the tipping point for humanitarian disasters, major population migrations, and conflicts over scarce resources. In many cases, these events could cause states to fail. The Pacific Command (PACOM) Area of Responsibility (AOR) is particularly vulnerable to these dangers. The Center for Naval Analysis Panel warned that climate change acts as a threat multiplier for instability in some of the most volatile regions of the world, with the potential to seriously exacerbate already marginal living conditions in many nations of the world.[7] Phenomenon likely will not remain isolated to small geographic areas or manifest themselves in just one environmental effect. The more likely scenario will involve multiple chronic conditions occurring across a broad geographic region. Consequently, the 2010 Joint Operating Environment analysis of future security lists climate change as one of the ten trends most likely to impact the joint force, joining other notable factors like globalization and energy.[8] For PACOM, clear implications arise; climate change events will challenge regional security, making an adaptive strategic posture versus a reactive one a strategic imperative.

A few basic premises concerning climate change appear relatively certain. Gradual warming has begun and will almost certainly continue, resulting in extreme weather events such as drought, flooding, sea level increases, retreating glaciers, habitat shifts, and the increased spread of life-threatening diseases. These impacts have great potential to create grave threats to future stability. While actions to reduce or eliminate the human causes of climate change remain vital, these actions face the biggest political hurdles and involve huge costs. Therefore, since preventative actions may not quickly mitigate the effects of climate change, adaptive strategies to prepare for inevitable challenges become critical for the United States to hedge against the instability that climate change may cause. Retired Marine Corps General and former commander of United States Central Command, Anthony Zinni, as part of the Center for Naval Analysis panel, warned in 2009 that failure to develop these strategies now will result in a price paid in military terms and human lives in the future.[9]

Climate Change as a Threat to National Security

Joshua Busby, a professor at the Lyndon B. Johnson School of Public Affairs at the University of Texas, has written extensively on the links between climate change and security for the Council on Foreign Relations and the United Nation's High Level Panel on Threats, Challenges and Change. He asserts that the short time horizons of decision makers in the United States pose a tough test for those seeking to make the case that climate change is a national security risk.[10] Extreme weather events with clear ties to climate change that instigate large scale destabilization have yet to occur. The effects of incremental warming of the atmosphere have occurred with no "headline event" driving American strategy or policy formulation. In addition, traditional views of national security involve protecting the state's territorial integrity or position on the world stage, and typically rely on violence or threat of violence as the means for response. However, for the United States, the realities of the post-WWII era expanded definitions of its national security to include a broader conception of defending interests around the world to enable the American way of life and promote its values. From protecting access to natural resources in the Middle East to preventing humanitarian disasters in the

Balkans, the United States' recent history reveals a willingness to fight those who threaten our expanding notion of national security. Busby includes climate change in this expanded definition of national security threats, proposing that vital interests are tied to the country's way of life and regarded as so important that a threat to them could be considered a challenge to national security.[11] The effects of climate change, resulting from catastrophic events like floods, or occurring persistently in the form of droughts or sea level rise, could easily disrupt the global economic and security environment. Without purposeful adaptive strategies to mitigate unacceptable risk, climate change poses a grave threat to American national security and vital interests.

American strategic adaptation to the results of climate change has already occurred by necessity. Seeking unity of effort in the Arctic, geographic combatant command lines of authority were re-drawn in the 2011 Unified Command Plan to designate Northern Command as the lead in the Arctic. Almost certainly, the summer sea ice in the Artic will be gone by the middle of the 21st Century.[12] This will open up new routes for shipping for parts of the year and provide a much shorter route between Europe and Asia. Disagreements over sovereign control of these waters and nebulous territorial boundaries that mattered little when the lines in question were ice could now foment international disputes.

Additionally, the elimination of summer sea ice opens access to vast, previously inaccessible petroleum reserves. In 2007, Russia upped the stakes in pursuit of these resources by laying claim to the North Pole and the resources underneath it. When the United Nations Security Council met in July of this year to consider whether climate change constituted a threat to international peace and security, Russia not surprisingly blocked adoption of strong language on climate change.[13] Their obstruction logically corresponds to their interests and aspirations in the Arctic and their economy's reliance on the sale of fossil fuels. A frustrated United States Ambassador to the United Nations, Susan Rice, called for the Council to recognize and tackle one of the "central threats to our age" and labeled Russia's intransigence "pathetic, shortsighted and a dereliction of duty."[14] Clearly, these tensions over the Arctic resulting from climate change offer a glimpse of future national security challenges, and demonstrate the necessity for a prioritization of adaptive climate-change strategies.

To a much greater degree than does the United States, the rest of the world considers climate change a grave threat to their interests. Possessing the world's largest economy, the United States is viewed as the de-facto leader of the industrial world. Global perception of "who's to blame" for the effects of climate change starts with the United States. According to Busby, a United States that fails to respond in the form of adaptive strategic assistance to less capable nations could find itself further alienated from poor countries who will seek to link any environmental catastrophe to climate change, valid or not.[15] He adds that demands for American humanitarian intervention would certainly increase in a world of fragile states buffeted by climate change, because a government's inadequate response to a natural disaster can undermine its legitimacy with its population.[16] Simply put, as climate change phenomenon threaten the sovereignty of weak states, America's inability or unwillingness to act in meaningful ways to deal with environmental impacts in those states could significantly diminish American influence.

While the world's public square perceives the United States as owing the most to developing adaptive strategies to mitigate climate change's impacts, it also views the United States as negligent in acknowledging climate change threats and in signing on to international efforts to combat it. In their extensive 2010 analysis of energy security in the 21st Century, Brookings Institution energy policy experts Pascual and Elkind describe a growing international consensus that a market mechanism such as cap-and-trade system or carbon tax should be at the heart of any effort to reduce greenhouse gas emissions.[17] While these policies seek climate change prevention versus adaption, they importantly demonstrate a growing disparity between American policy on climate change and what most of the rest of the world desires.

The United States has resisted international efforts to implement policies to reduce the greenhouse gas emissions blamed for climate change. The Kyoto Protocol, which attempted to put in place binding rules on emissions and was not signed by the United States, expires in 2012. Despite clear concerns with the economic impacts to the American economy if it had signed on, the United States holds a position clearly in the minority. Seeking to temper these negative perceptions, the United States led efforts in 2009 to develop the Copenhagen Accord through the auspices of the United Nations to recognize climate change as a threat and demonstrate the political will to develop strategies to combat it. The Pew Center on Global Climate Change summarized the impact of the Copenhagen Accords by characterizing them as a "political (as opposed to legal) agreement of a novel form," showing its frustration with the Accords' lack of binding commitments or enforcement mechanisms.[18] Pasqual and Elkind explain the United States' reluctance to sign on to international climate change initiatives with teeth by asserting that America remains resistant to ceding sovereignty over energy—and by extension national security—to external, international intervention.[19]

This all adds up to a national security threat that at first may seem farfetched but becomes increasingly more plausible as its multiple effects become apparent. Drastic climate change effects could lead to American military involvement defending forests in Indonesia or Brazil, pursuing environmental bad actors as part of efforts to enforce new international mandates on climate change, or defending American interests against an increasingly hostile world that places blame on the United States for climate change impacts. Even al-Qaeda leaders cite global warming repeatedly in propaganda intended to foment anti-American sentiment.[20]

The national security threat presented by climate change thus demands adaptive strategies now to deal with inevitable consequences of climate change in an effort to balance probable American resistance to adopting climate change prevention policies. In its 2010 report on climate change and United States Armed Forces, the Center for New American Security (CNAS) concluded that the effects of climate change are likely to reshape the current and future security environment.[21] Simply put, while the comprehensive national security dilemma presented by climate change encompasses difficult policy decisions across the spectrum of diplomatic, informational, military and economic arenas, adaptive strategies offer a clearer path towards positioning the United States as a leader in blunting potential impacts than do preventive strategies. Adaptive strategies have important strategic implications due to the likelihood that the breadth of the challenges presented by climate change will preclude the United States from helping everyone. Momentum for proper strategic planning increasingly trends toward recommendations like the CNAS' 2010 assertion that Combatant Commands must develop strategies to help partner countries adapt in ways that will hedge against destabilizing forces inherent in climate change.[22] The implication: a gap exists between current Combatant Command capabilities to address the regional and security challenges presented by climate change phenomenon and the necessary development of a multi-national adaptive strategic alliance to effectively combat the phenomenon.

American Strategic Guidance Concerning Climate Change

Fortunately for Combatant Commanders responsible for executing adaptive strategies in their geographic areas of responsibility, clear national policy on climate change exists. This is a relatively recent development. During Congressional budget negotiations in 2008, the DoD had to include climate issues in its strategic plans to facilitate budget approval. When the Obama Administration entered the White House, it made climate change policy a key part of strategic planning, bringing it forward from segmented, scientific debate to what the New York Times characterized as "a central policy focus."[23]

The 2010 National Security Strategy (NSS) provides extensive strategic vision and national policy on climate change. Like the CNAS assessment, it calls for fostering regional alliances to combat climate change so countries can adapt and mitigate its impacts.[24] Building on this guidance for strategy development, the 2010 Quadrennial Defense Review (QDR) lists climate change as one of four specific issues imperative for Defense Department reform.[25] It proposes that climate change will shape the strategic environment and the roles and missions that the United States military will undertake. The QDR calls for proactive engagement with vulnerable countries to help build their capability to respond to climate change challenges and events.[26] Simply put, the NSS and QDR direct strategic emphasis on regional adaptive strategies for inevitable climate change challenges.

Surprisingly, the 2011 National Military Strategy (NMS) barely mentions climate change, stating only that its uncertain impact may challenge the ability of weak or developing states to respond to natural disasters.[27] Similarly, the Chairman of the Joint Chief's Capstone Concept for Joint Operations describes an operational need to act in concert with international partners and calls for military forces to begin partnering with other federal agencies and local authorities for relief and reconstruction efforts.[28] It does not however direct leaders specifically to consider climate change impacts to meet these cooperative requirements. Strategic military guidance appears misaligned with national policy on climate change.

In his 2011 congressional testimony to lay out strategic priorities, the Commander of United States Pacific Command prescribed force posture aligned to meet emerging 21st Century threats and highlighted his region's vulnerability to natural disasters.[29] The testimony acknowledged a need for significant assistance from the international community to respond to environmental challenges in South Asia, but did not acknowledge the increasing frequency and severity of these events due to climate change. In contrast, Pumphrey concluded in her extensive look at the national security implications of climate change that a regional approach to adaptive strategies led by combatant commanders offered the most feasible and reasonable strategy to combat inevitable climate change challenges.[30] At the combatant command level where vision and policy should meet actionable strategy, adaptive strategies to handle climate change events appear to be underrepresented.

Simple reasons exist for the disconnect between national policies and actionable strategies. First, as previously addressed, most strategic discussion of climate change focuses on hard choices to reduce greenhouse gas emissions at the expense of aggressively addressing adaptive strategies aimed at inevitable climate change effects. From the DoD perspective, this has meant a focus on reducing energy consumption and emissions at facilities. These actions fall under the "organize, train, and equip" responsibilities assigned to services, not Combatant Commander's Theater Strategies.

Second, national policy clearly places the DoD in a supportive role for addressing climate change. The State Department's first Quadrennial Diplomacy and Development Review, a partner document to the QDR, prescribed the State Department as the leader for interagency delegations to conferences on climate change.[31] However, much of the State Department's strategy on climate change focuses on negotiating international agreements on greenhouse gas emission policies aimed at preventing climate change with no significant inclusion of adaptive strategies.

Finally, like the other geographic combatant commands, PACOM adroitly handled recent humanitarian assistance requirements in its area of responsibility. From the tsunamis that decimated Thailand and Indonesia in 2004 to the recent earthquake and tsunami responses in Japan, military forces undertaking their humanitarian assistance and disaster response missions in the Pacific have performed exceptionally, saving lives, promoting stability, and furthering American interests. At question is whether this same reactive posture can continue to succeed if climate change ratchets up the intensity, frequency and scope of natural disasters, or whether a greater emphasis on adaptive planning is required.

In 2009, the DoD published DoD Instruction 3000.05 "Stability Operations" in response to growing recognition that traditional roles and missions had to evolve to deal with 21st Century realities. Defining stability operations as a primary mission of the military, it prescribed military activities synchronized with other instruments of national power to establish security, provide essential services, reconstruction, and humanitarian relief.[32] It tasked Combatant Commanders with incorporating stability operations activities and concepts into training, exercises, and most importantly planning.[33]

In September of 2011, the Joint Staff released Joint Publication 3-07, providing extensive guidance on stability operations to complement DoD Instruction 3000.05. It defines stability operations as the various missions, tasks, and activities conducted outside the United States in coordination with other instruments of national power to accomplish four objectives: maintain or reestablish a safe and secure environment, provide essential governmental services, accomplish emergency infrastructure reconstruction, and provide humanitarian relief.[34] It further directs joint commanders to evolve integrated planning processes to include the wider international community, host nations, and other multinational partners.[35] While this mission rose to prominence from the insurgent conflicts in Iraq and Afghanistan, it would apply perfectly to a necessary shift in PACOM's strategy in the direction of adaptive climate-change risk mitigation. Climate change events can easily threaten the governments of fragile states and drive the United States to undertake stability operations in those nations. Nation-building objectives utilizing stability operations doctrine become necessary not just in states where the United States forces regime change, but also in weak or failing states that may come apart at the seams due to natural disasters resulting from climate change.

Fortunately, joint doctrine well defines the types of missions likely undertaken by the United States military as part of a heightened priority on adaptive climate change strategies. Whether as part of stabilization efforts in the Middle Eastern war zones, regional theater engagement exercises, or recent disaster relief missions, American military forces maintain skills of the exact type demanded by climate change phenomenon. First, climate change events could easily result in an increased frequency of non-combatant evacuation operations (NEO). Second, responses to environmental disasters caused by climate change include what Joint Publication 3-07 describes as programs to meet basic human needs to ensure the social well-being of the population.[36] Whether providing the lift capability to move large numbers of people and equipment in all domains, undertaking large-scale medical response, delivering and distributing relief supplies, or providing security, the United States military stands out as the only United States Government entity with the capacity and funding to undertake meaningful, timely actions to address climate change phenomenon.

Thus, adopting a strategic shift towards higher prioritization of adaptive strategies would require no expansion of military doctrine or roles and missions of military forces. Instead, Combatant Command focus on adaptive strategies would ensure that national policy priorities on climate change translate into feasible, necessary plans to deal with inevitable climate change phenomenon. Simply put, the necessity of this strategic shift derives from a perception that shortfalls exist in risk analysis and mitigation strategies for climate change in PACOM's theater strategy. Shoring up this gap would take PACOM's climate change posture from reactive to proactive.

The Geopolitical Context of the Pacific

In his 2011 congressional testimony, Admiral Robert Willard described his area of responsibility as vital to American national interests, spanning half the earth, home to more than three billion people, the world's three largest economies, and one-third of the United States' annual trade value.[37] He listed eight challenges to sustaining the conditions that underpin prosperity in PACOM. In addition to traditional military threats such as China's rise and North Korea's isolation and nuclear ambitions, his testimony listed humanitarian crisis and natural disasters as asymmetric challenges driving the need for forward presence.[38] While China and North Korea naturally dominate much of PACOM's strategic planning, American military activities in the last ten years in the region have primarily involved humanitarian crisis response to natural disasters.

Water dominates the geopolitical context in the Pacific. Approximately forty percent of the population, roughly four billion people, live within forty five miles of the coast.[39] Rising sea levels and weather events characterized by their increased frequency and intensity as a result of climate change pose clear humanitarian disaster threats. As PACOM, in conjunction with interagency partners, seeks to build relationships in the region that further American interests, climate change increasingly becomes a significant factor that will shape relationships. Where PACOM worries about stability vulnerabilities of young democratic governments and weak states, climate change phenomenon threaten those states to an increasing degree. Where PACOM has concerns with regional competitors, primarily China, actions to combat climate change play an increasingly significant role in the informational campaign for influence in the region.

Some climate change vulnerabilities in the Pacific clearly present themselves. Many countries in Asia rely on glaciers in the Tibetan plateau for their drinking water. If the glaciers vanish, as trends indicate, the resultant droughts and battles over dwindling water resources could easily result in state-on-state conflict. In September of 2011 New Zealand hosted the annual Pacific Islands Forum. The three top issues dominating the conference: climate change, political instability, and economic reforms. United Nations Secretary-General Ban Ki-moon attended the conference and stated that climate change threatens the very survival of Pacific Island countries.[40] Kiribati's president reported rising sea levels have forced his nation to consider moving its one hundred thousand-strong population onto man-made floating islands.[41] Bangladesh, with a population of one hundred fifty million, faces similar vulnerabilities to sea level rise. Ten percent of its population lives within three feet of sea level. Flooding could easily drive mass migration to India, a concern so significant that India has begun building a fence to prevent such an "invasion."[42] Clearly, sea level rise represents a major concern in the region and threatens stability.

Indonesia provides the best example of a vital PACOM state in which all of the conditions exist for climate change to create instability that could dangerously challenge American interests. Indonesia represents the world's fourth most populous nation, third largest democracy, and largest Muslim-majority country. In his congressional testimony, Admiral Willard described it as an emerging vibrant democracy with an increasing leadership role in Southeast Asia with which PACOM has broadened its military engagements.[43] Clearly, a solid relationship with a stable Indonesia supports American interests in Southeast Asia. While Indonesia's democratic reforms are encouraging, it remains a very young, relatively fragile democracy. Terrorist groups there seek opportunities to challenge the democratic movements. Busby asserts that, because of its vast forests, Indonesia plays an important role in determining impacts of climate change. Deforestation and fires recently made Indonesia the world's third largest greenhouse gas contributor behind the United States and China. As with the rain forests in South America, Indonesia's remaining forests also represent a major reducer of greenhouse gases as they filter carbon from the air and store it, leading Busby to predict increasing pressure to pay Indonesia to protect them.[44] Thus whether its forests become pawns in the greater climate change debate or whether sea level rise or other natural disasters spurred by climate change threaten Indonesia's stability and American access to sea lanes and natural resources, PACOM must carefully examine its role in developing adaptive strategies to successfully counter climate change events in Indonesia.

China represents a different challenge altogether. Most American engagement with China on climate change has focused on it being a major offender in increased greenhouse gas emissions and thus has focused on preventative climate change strategies. While these efforts are vitally important, China stands out in PACOM as the place where successful adaptive strategies present the clearest opportunities to further American interests. Secretary of State Clinton recently described the next hundred years as "America's Pacific Century," prescribing adaptive regional alliances that address new challenges and opportunities, but also operationally and materially can handle the full spectrum of state and non-state threats.[45] As the United States' strategic focus shifts toward the PACOM region, engagement with China increasingly presents a challenging dilemma. China's portion of the Pacific Rim economy has grown significantly over the last decade and will continue to expand. At question is whether this will lead to violent confrontation with the United States or whether the global economy can accommodate China's rise without conflict. One thing is certain—regional stability is a prerequisite for any scenario leading to peaceful coexistence. Thus a PACOM regional focus that prioritizes and develops adaptive climate change strategies will greatly increase the likelihood of stability and a mutually beneficial relationship between the two superpowers. A shift in the focus of PACOM's military planning emphasis away from large-scale kinetic conflict and instead centered on aiding nations to combat climate change events would be viewed positively by China and decelerate trends towards great power conflict in the Pacific. This shift in priority would not exceed acceptable risk thresholds by ignoring the unlikely but catastrophic possibility of major theater war, but would likely provide exponential benefits to adaptive climate change engagement.

Admiral Willard's 2011 Congressional testimony included, as expected, much discussion on China; however, while much of the testimony addressed engagement with China, it did not include climate change strategy as an engagement opportunity. Arguably the absence of dialogue on climate change opportunities with China appropriately supports the "other interagency lead" for climate change strategy. It does however neglect potential adaptive engagement that only PACOM can undertake. Whether China's public position recognizes climate change as a grave threat to its own security or whether it seeks to use the issue as leverage to promote its interests in the region, China's policy towards addressing climate change as a serious threat has evolved rapidly over the past few years. At the completion of the Copenhagen Accord meetings in 2009, China's prime minister said that to meet the climate change challenge, the international community must strengthen confidence, build consensus, make vigorous efforts and enhance cooperation.[46] China's chief spokesperson on climate change described it as a more serious issue for China than even the global economic crisis.[47] China appears resolute in its efforts to aggressively address climate change and threatens to take the upper hand on this strategic narrative in the court of world opinion.

In *Vacuum Wars*, a groundbreaking analysis of the 21st Century challenges created by failed states, Jakub Grygel warns that the vacuum created by weak or collapsing states provides areas for great power competition, stating that the interest of great powers is often not to rebuild the state for humanitarian purposes themselves, but to establish a foothold in the region, obtain favorable economic deals, and ultimately to weaken the other great power.[48] China's desire to increase its sphere of influence can be characterized not as a desire to spread an ideology, but to gain access to natural resources. Humanitarian relief missions to support PACOM nations buffeted by climate change phenomenon offer a clear opportunity for China to further its interests with those nations. Combining these efforts with an information campaign portraying actions as greater than those undertaken by the United States to reduce greenhouse gas emissions at home has the potential to ingratiate China with Pacific nations and supplant the United States as the standard bearer for combating climate change.

According to an April, 2010 *Foreign Policy* article, China's recent commission of a hospital ship and identification of disaster relief as a key mission for a future aircraft carrier foreshadow Chinese military assets delivering Chinese-made disaster-relief supplies in the not too distant future.[49] Thus all signs point to evolving Chinese climate change strategy that addresses both prevention and adaptive strategies that address regional concerns. As a result of China's focus on climate change strategy, the United States faces erosion in its desired narrative as both the champion of humanitarian assistance and disaster relief in the Pacific region, and the mantle of the nation most prepared to handle climate change phenomenon in PACOM. For the United States, development of adaptive strategies that coexist with and even complement Chinese efforts offer an path towards mitigating climate change impacts for both humanitarian and national-interest reasons while simultaneously shaping a mutually beneficial relationship instead of a costly rivalry with this rising peer.

Recommendations for PACOM Adaptive Strategies

Busby predicts the demands for humanitarian intervention will likely increase in a world of fragile states buffeted by climate change.[50] Climate change clearly represents a security challenge for PACOM and requires adaptive strategies to address it. PACOM has risen admirably to disaster relief demands in the 21st Century, but the increasing severity and frequency of climatic phenomenon propelled by climate change demand greater strategic planning prioritization. This adaptive strategy should focus on three strategic tenets—promulgating a strategic narrative that promotes a positive perception of the United States, developing a cooperative coalition among as many PACOM states as possible addressing climate change phenomenon as its central focus, and undertaking a vulnerability analysis of climate change events for PACOM states.

In forging a successful strategic narrative on climate change in the Pacific, PACOM must focus its message on recognition that environmental climate change phenomenon present grave threats to the region, and that PACOM will aggressively pursue adaptive strategies with all states seeking to lessen the impacts of climate change events. Two 21st Century environmental disasters in the Pacific teach important lessons on the opportunities relating to a clear, focused American strategic narrative. While climate change itself did not cause either one, the lessons taken from them directly correlate to the need for adaptive climate change strategy.

In 2004, a series of tsunami waves devastated coastal areas in Indonesia and Thailand. At that time, China lacked the capabilities and strategic intent to provide humanitarian assistance. Focused on the wars in Iraq and Afghanistan and because the disaster occurred during Christmas, the United States reacted slowly. However, with PACOM in the lead, the United States eventually developed a humanitarian assistance effort facilitated almost exclusively by American military capabilities. The current positively-trending American-Indonesian relationship previously described had its genesis in the Indonesian reaction to these American humanitarian efforts. Averting an even greater humanitarian disaster dampened potential instability that could have threatened the government in Jakarta. Realist and liberal strategic thinkers welcomed the positive impact towards American interests and the application of American military might for humanitarian purposes. American efforts helped avert a far greater humanitarian disaster, and facilitated closer relationships and greater cooperation on other regional issues important to Unites States national security.

Similarly, the 2011 earthquake and tsunami in Japan may hold a valuable lesson on humanitarian assistance and disaster relief and how it can influence global perceptions of the United States. Recent national surveys in Japan report a record eighty two percent of Japanese have friendly feelings toward the United States, perceptions springing from the good will created by Operation Tomodachi and insecurities over China's rise.[51] While many Japanese would likely have positive views of the United States without the Tomodachi effort, the operation critically influenced Japanese perceptions to positive levels higher than at any time since polling began in 1978.

America's relationship with Japan has increased in significance with China's rise. In an interview where Admiral Willard labeled the catastrophe the most complex disaster environment he'd experienced in his thirty seven year career, a Japanese journalist asked him to assess the impact of the American-Japanese alliance towards efforts to rapidly respond to the disaster. PACOM's Commander responded by emphasizing the importance of regularly conducting disaster response exercises.[52] When PACOM undertakes these missions with the benefit of recently accomplished exercises promoting interoperable partnerships, the capability to respond effectively rises exponentially. PACOM exercises like Pacific Angel in 2011, which provided a week of extensive medical, engineering, and infectious disease control in Indonesia, represent baseline engagements that should be expanded to as many states in the region as possible.[53]

To deal with the emerging challenges of climate change phenomenon in the PACOM region, promulgating a strategic narrative that promotes a positive perception of the United States starts with a strategic focus on humanitarian assistance and disaster relief. No other military mission provides the same opportunities to further

American interests at such a low cost, and no other United States agency can address all of the challenges presented by humanitarian assistance and disaster relief.

Since humanitarian assistance efforts make so much sense and do so much to promote a positive regional perception of the United States, a cooperative regional strategy that builds upon these foundations would provide tremendous opportunities for PACOM. Herein lies the crux of the argument toward an increased focus on adaptive response to climate change and forms the nexus where ideas meet actionable strategy. According to Pumphrey, climate change can only be effectively addressed by cooperation and thus offers the United States an opportunity and rationale to foster partnerships and build trust.[54] A PACOM regional alliance based primarily on collective capabilities to combat inevitable climate change environmental phenomenon would both mitigate risk and provide common cause. Certainly the preponderance of assets allowing rapid adaptive responses to events would be provided by the United States military, but even states with small economic means could focus on critical, unique assets and mission contributions. The alliance's proactive posture towards climate change events would be maximized by the sum of its collective stakeholders, forging partnerships against a common enemy. Regional enthusiasm for this type of alliance would likely be high and bridge some of the perceived gap between the United States and the rest of the region on the threats inherent in climate change. Most importantly, a climate change alliance would offer a clear opportunity to engage militarily with China in a positive direction. Critical in today's economically constrained budget environment, a regional military team poised to address climate change environmental phenomenon would not require significant change in PACOM's force structure or infrastructure.

PACOM's risk imbalance in its posture to deal with climate change results not from shortfalls in force structure, but rather from gaps in strategic planning and focus.

Movement toward a regional alliance focused on climate change would necessarily drive needed interagency cooperation as well. Presidential Directive-44, Management of Interagency Efforts Concerning Reconstruction and Stabilization, designated the State Department as the lead to coordinate stabilization efforts, directed the State Department and DoD to work together in this effort, and created a working group to coordinate and synchronize efforts.[55] Thus the political infrastructure exists for PACOM to engage with the interagency to push a more integrated regional and whole-of-government posture to deal with climate change in its region. Adding a greater interagency aspect to a regional climate change alliance would only add to its legitimacy and effectiveness.

Another historical example shows the need for a regional alliance focused on adapting to climate change to manage the risks to stability inherent in climate change. America's response to Hurricane Katrina exposed extensive weaknesses in our own government's ability to react quickly to a natural disaster and coordinate a cohesive whole-of-government response. If the United States found itself significantly unprepared for the challenges brought by Katrina, a similar event could likely challenge Pacific nations to an even greater extent. Much to its surprise, the United States received significant international support to handle Katrina's challenges. The types of challenges predicted to result from climate change phenomenon will likely exceed a purely American ability to respond. A regional alliance offers greater overall capability and cost sharing than a unilateral American response. The new stability joint doctrine prescribes the inclusion of a wider international community, including host nations and nongovernmental organizations in integrated planning processes to deal with regional challenges.[56] Simply put, to promote a posture on climate change that maximizes American interests in the Pacific, PACOM advocacy for creating a regional alliance to adapt to and combat inevitable climate change is needed.

Finally, a strategic shift in focus towards adaptive strategies in PACOM requires an extensive vulnerabilities analysis. Katrina again offers a salient lesson in this regard. In many ways, the United States was unaware of the extent of the vulnerabilities in New Orleans, and did not know what to do when the levees failed. The disjointed response to Katrina resulted from a lack of strategic priority to understand in advance the risks and vulnerabilities of the region to a large hurricane. Fortunately, America seems to have learned its lesson domestically, as evidenced by the extensive planning and preparation for the storms that threatened the Northeast and New York City in 2011. A similar in-depth risk analysis applied in PACOM would mitigate risk and uncertainty. Pumphrey warns that we stand to lose a great deal if we do not move fast because the evidence suggests that the problems will only multiply if we wait.[57]

Like all geographic combatant commands, PACOM prioritizes engagement with regional states based on national interest. Adaptive climate change strategy would meld these priorities with a thorough understanding of the unique vulnerabilities to climate change phenomenon for each of these states. Busby strongly asserts that a better understanding of vulnerabilities to climate change is needed, stating that the analysis will require more work by the physical sciences to identify which places are most vulnerable.[58] Clearly, figuring out what climate change events most threaten a PACOM nation and the extent of the threat lie well outside traditional military expertise and missions. However, since PACOM would likely have to respond to these events if they occur, its role in facilitating vulnerability studies requiring extensive non-military expertise logically arises. In the end, a PACOM-led climate change vulnerability assessment would require extensive involvement from interagency and non-governmental experts as well as experts from the subject nation. PACOM's Joint Interagency Coordination Group (JIACG) would seem a logical focal point to merge non-military inputs into country desk officer's humanitarian assistance and disaster-relief planning.

A three-pronged approach towards effective strategic narrative, a regional alliance, and climate change vulnerability risk assessment would result in an adaptive theater strategy that mitigates the risk of climate change driving instability in the region that could result in disastrous conflict and human tragedy. By using existing military planning and design concepts such as "most likely" and "most dangerous" analysis of climate change in the region, PACOM would position itself as well as possible to deal with climate change and further American interests. This construct could obviously serve as a template for the other geographic combatant commands.

Conclusions

Some clear conclusions present themselves concerning the climate change threat for PACOM. Climate change undoubtedly has the potential to produce tremendous instability in the region. While the United States' national security policy makes climate change a top priority, strategy formulation at the combatant command level appears to fall short of needed actions to translate policy into regional strategies that adequately address risks inherent in likely climate change events. The problem goes beyond acknowledgment of needed proficiency in humanitarian assistance and disaster relief missions in PACOM's Theater Strategy. A simple shift in strategic priority to assess vulnerabilities and develop a regional alliance focused on adaptive strategies for climate change would address shortfalls. Multi-national exercises of the scenarios identified by the vulnerability analysis and alliance relationships are vital, and would not necessarily come at the expense of exercises to hedge against the traditional military threats presented by China and North Korea. American national policy, military doctrine, and interagency guidance clearly prescribe a need for adaptive climate change strategy now to hedge against likely environmental events that will challenge America's interests.

Proactive, adaptive strategies require significant up-front planning efforts and commitment, but also offer tremendous payoffs in engagement opportunities and strategic narrative. These recommendations meet feasibility criteria because they require a minimal shift in strategic priority and would be welcomed by a region with far greater concern for climate change than currently demonstrated by the United States. The recommendations pass measures of acceptability as well because they clearly fall in line with national guidance and policy and are very inexpensive to undertake since existing shortfalls lie not in force structure but in strategic plans. Finally, when judged on merits of suitability, a regionally focused adaptive climate change strategy correctly shifts priority towards prudent risk mitigation of unacceptable vulnerabilities. In making similar recommendations, Busby reminds us that even if the consequences of climate change prove less severe than feared, adaptive climate change strategies that result in improved humanitarian assistance and disaster relief capabilities have exponential benefits.[59] While the gap in current PACOM views on climate change and the requirements to implement these initiatives appears relatively narrow, a mostly reactive posture towards climate change phenomenon carries significant risk. The time to implement these changes is now.

Endnotes

[1] CNA Corporation, *National Security and the Threat of Climate Change* (Alexandria, VA: CNA Corporation, April 30, 2007), 56.

[2] These include the National Academy of Sciences, National Oceanic and Atmospheric Administration, National Air and Space Administration, and the United Kingdom's Hadley Centre for Climate Change.

[3] Ibid.

[4] Gregory F. Treverton, *Making Policy in the Shadow of the Future* (Santa Monica, CA: RAND Corporation, 2010), x.

[5] Carolyn Pumphrey, *Global Climate Change: National Security Implications* (Carlisle, PA: U.S. Army War College Strategic Studies Institute Press, 2008), 13.

[6] CNA Corporation, *National Security and the Threat of Climate Change,* 6.

[7] Ibid.

[8] General J.N. Mattis, USMC, *Joint Operating Environment, 2010* (Suffolk, VA: U.S. Joint Forces Command, February 2010), 32.

[9] John M. Broder, "Climate Change Seen as Threat to U.S. Security," August 9, 2009, http://www.nytimes.com/2009/08/09science/earthe/09climate.html (accessed August 22, 2011).

[10] Joshua W. Busby, "Who Cares about the Weather?: Climate Change and U.S. National Security, *Security Studies,*" 17:3, 473.

[11] Ibid., 475.

[12] Joshua W. Busby, *Climate Change and National Security: An Agenda for Action* (Washington, D.C.: Council on Foreign Relations, 2007), 7.

[13] Bill Hewitt, "Climate Change = Security Threat," July 21, 2011, http://foreignpolicyblogs.com/2011/07/21/climate-change-security-threat/ (accessed September 22, 2011).

[14] Ibid.

[15] Busby, *Who Cares about the Weather?: Climate Change and U.S. National Security,* 503.

[16] Ibid., 498.

[17] Carlos Pascual and Jonathan Elkind, *Energy Security: Economics, Politics, Strategies, and Implications* (Washington D.C.: Brookings Institution, 2010), 223.

[18] Elliot Diringer, "Summary: Copenhagen Climate Summit," http://pewclimate.org/international/copenhagen-climate-summit-summary (accessed September 22, 2011).

[19] Pascual and Elkind, *Energy Security: Economics, Politics, Strategies, and Implications*, 254.

[20] Pew Center on Global Climate Change, "National Security Implications of Global Climage Change," www.pewclimate.org (accessed September 22, 2011).

[21] Commander Herbert E. Carmen, USN, Christine Parthemore, and Will Rogers, *Broadening Horizons: Climate Change and the U.S. Armed Forces* (Washington, D.C.: Center for a New American Security, 2010), 1.

[22] Ibid., 3.

[23] John M. Broder, "Climate Change Seen as Threat to U.S. Security," August 9, 2009, http://www.nytimes.com/2009/08/09science/earthe/09climate.html (accessed August 22, 2011).

[24] Barak H. Obama, *National Security Strategy* (Washington D.C.: The White House, May 2010), 47.

[25] Robert M. Gates, *Quadrennial Defense Review Report* (Washington D.C.: U.S. Department of Defense, February 2010), 73.

[26] Ibid., 84-85.

[27] Michael G. Mullen, *The National Military Strategy of the United States of America* (Washington D.C.: Chairman of the Joint Chiefs of Staff, February 2011), 2.

[28] Michael G. Mullen, *Capstone Concept for Joint Operations* (Washington, D.C.: Chairman of the Joint Chiefs of Staff, January 2009), 19.

[29] Robert F. Willard, *Statement of Admiral Robert F. Willard, U.S. Navy, Commander, U.S. Pacific Command*, Before the House Appropriations Committeee Subcommittee on Defense on U.S. Pacific Command Posture (Washington, D.C.: House Appropriations Committee Subcommittee on Defense, April 2011), 5, 19.

[30] Pumphrey, *Global Climate Change: National Security Implications*, 177.

[31] Hillary R. Clinton, *Leading Through Civilian Power: The First Quadrennial Diplomacy and Development Review* (Washington, D.C.: U.S. State Department, 2010), 2.

[32] U.S. Department of Defense, DoD Instruction 3000.05 (Washington D.C.: U.S. Department of Defense, September 16, 2009), 1.

[33] Ibid., 14.

[34] Vice Admiral William E. Gortney, USN, *Joint Publication 3-07: Stability Operations* (Washington D.C.: Director, Joint Staff, September 29, 2011), vii.

[35] Ibid., xiii.

[36] Ibid., xix.

[37] Robert F. Willard, *Statement of Admiral Robert F. Willard, U.S. Navy, Commander, U.S. Pacific Command*, Before the House Appropriations Committeee Subcommittee on Defense on U.S. Pacific Command Posture (Washington, D.C.: House Appropriations Committee Subcommittee on Defense, April 2011), 2.

[38] Ibid., 3.

[39] CNA Corporation, *National Security and the Threat of Climate Change* (Alexandria, VA: CNA Corporation, April 30, 2007), 24.

[40] Phil Mercer, "Climate Change Concerns Dominate Pacific Bloc Summit," September 7, 2011, http://www.voanews.com/english/news/asia/east-pacific/Fiji-Climate-Change-Concerns-Dominate-Pacific-Bloc-Summit-129368308.html (accessed September 7, 2011).

[41] Ibid.

[42] CNA Corporation, *National Security and the Threat of Climate Change* (Alexandria, VA: CNA Corporation, April 30, 2007), 24.

[43] Robert F. Willard, *Statement of Admiral Robert F. Willard, U.S. Navy, Commander, U.S. Pacific Command*, Before the House Appropriations Committeee Subcommittee on Defense on U.S. Pacific Command Posture (Washington, D.C.: House Appropriations Committee Subcommittee on Defense, April 2011), 15.

[44] Busby, *Climate Change and National Security: An Agenda for Action*, 20.

[45] Hillary Rodham Clinton, "America's Pacific Century," U.S. Department of State and *Foreign Policy Magazine*, http://www.state.gove/secretary/rm/2011/10/175215.htm (accessed November 8, 2011).

[46] "Copenhagen Deal Reaction," December 19, 2009, BBC News, http://news.bbc.co.uk/2/hi/science/nature/8421910.stm (accessed September 22, 2011).

[47] Rosemary Foot, "China and the United States: Between Cold and Warm Peace," *Survival*, 51:6, 128.

[48] Jakub Grygiel, "Vacuum Wars: The Coming Competition Over Failed States," *The American Interest*, July/August 2009, 3.

[49] Drew Thompson, "Think Again China's Military," *Foreign Policy*, March/April 2010, 3.

[50] Busby, "Who Cares about the Weather?: Climate Change and U.S. National Security," 498.

[51] "Japan Poll Finds Record Good Will for U.S.," December 5, 2011, http://214.14.134.30/ebird2/ebfiles/e2011205857177.html (accessed December 5, 2011).

[52] *United States Pacific Command Headlines Page,* http://www.pacom.mil/ (accessed September 7, 2011).

[53] "Pacific Angel 2011," Air Force Print News, http://www.pacaf.af.mil/news/story_print.asp?id=123259853 (accessed September 7, 2011).

[54] Pumphrey, *Global Climate Change: National Security Implications,* 12.

[55] United States Army War College, *Campaign Planning Handbook, Academic Year 2012* (Carlisle Barracks, PA: Department of Military Strategy, Planning, and Operations, AY 2012), 12.

[56] Vice Admiral William E. Gortney, USN, *Joint Publication 3-07: Stability Operations* (Washington D.C.: Director, Joint Staff, September 29, 2011), vii.

[57] Pumphrey, *Global Climate Change: National Security Implications,* 17.

[58] Busby, "Who Cares about the Weather?: Climate Change and U.S. National Security," 500.

[59] Ibid., 501.

www.ingramcontent.com/pod-product-compliance
Lightning Source LLC
Chambersburg PA
CBHW081812170526
45167CB00008B/3404